A New Frontier
Space Tourism and Private Space Exploration

Table of Contents

Chapter 1. Introduction

As humanity gazes beyond the stars, a lofty aspiration becomes an exciting new reality in our Special Report: "A New Frontier: Space Tourism and Private Space Exploration". Step into a world where captivating breakthroughs in technology blend with the untamed beauty of the cosmos. This report unravels the intricate tapestry of private space exploration, with a grounded, digestible analysis of its complex engineering undertakings and the emergence of a thrilling new market - Space Tourism. Indulge your curiosity and ignite your imagination as this report unravels the potential of the newest frontier. The promise of space, once limited to astronauts and the elite, is becoming more accessible, sparking a revolution that will change our perspective on life, our planet, and the endless universe beyond. Dive into this Special Report and get your ticket to the stars. The future is closer than you think.

Chapter 2. The Dawn of Private Space Exploration

In the early 2000s, the pioneering spirits of the new millennium turned their gazes skyward, planting the seeds of private space exploration. The foundations of this technological leap were laid not just by vigorous competition but ignited by a shared dream of awakening a new dawn, where space was no longer the domain of state institutions alone.

2.1. Early Initiatives and Influential Companies

Seeds of private space exploration were sown in the form of pivotal competitions, like the 'Ansari X Prize.' This competition awarded $10 million to the first non-government organization that could launch a reusable manned spacecraft into space twice within two weeks. Winner SpaceShipOne, funded by billionaire engineer Paul Allen and designed by Burt Rutan, became an inspiration for many, signifying that space was not just a playground for governmental behemoths but an arena where private enterprise could thrive.

Another early facilitator of private launches was SpaceX, an ambitious vision of Elon Musk. Launched in 2002, SpaceX has remained at the forefront of technological breakthroughs, with creations like the Falcon 1, Falcon 9, Falcon Heavy, and most recently, Starship. Their approach towards reusable rockets disrupted the economics of space exploration by significantly reducing the costs. SpaceX's advent marked the dawn of a new era where private enterprises could not only dream but actively reshape the reality of space exploration.

2.2. Innovation and Expansion

In due course, multiple companies began to venture into space exploration, proving the viability of a private approach. Companies like Blue Origin, led by Amazon's founder Jeff Bezos, Virgin Galactic from tycoon Richard Branson, and Orbital Sciences, now Northrop Grumman, expanded their engineering and technological horizons by delving into building rockets, spaceplanes, and satellites. These organizations, each unique in their strategies, contributed to the diversification of the industry, making space exploration a comprehensive venture rather than remaining as simple rocket launches.

The implications of these developments were far-reaching. As private companies furthered their abilities, traditional governmental institutions like NASA began investing in partnerships with these firms, aiding in the expansion of the space frontier. In a remarkable shift of strategy, NASA outsourced certain sectors like cargo resupply missions to the International Space Station (ISS) to commercial entities, an endeavor successfully carried out by SpaceX and Northrop Grumman.

Moreover, NASA's Commercial Crew Program, aimed to develop a new spacecraft and launch system to ferry crew to the ISS and back, saw active competition among private entities. This initiative exemplified the large role private companies began playing in shaping the future of space exploration and travel.

2.3. Technological Advancements

The advent of private entities into the fold of space exploration led to a string of technological advancements that were profound and, at many times, disruptively innovative. SpaceX, notably, pioneered the technology of reusable rockets. Their Falcon rockets land vertically after delivering payloads to space, a feat that recycles hardware and slashes mission costs.

Newer companies, such as Rocket Lab, contributed to the development of small satellite launching technology, lowering the entry bar for businesses aiming to leverage satellite-based services. Equally noteworthy is the development of suborbital spaceplanes by Virgin Galactic and Blue Origin, handling space tourism initiatives that promise exhilarating (if brief) experiences of weightlessness to anyone who can afford a ticket.

2.4. Future Horizons

In this era of private space exploration, one clear theme resonates – there are no limits. Companies are now expanding their ambitions beyond simple launches, looking at big-picture objectives as diverse as lunar tourism, asteroid mining for resources, constructing hotels in space, and even planning colonies on Mars. This change in narrative, from a space race between nations to a collaborative effort between public and private sectors, is redefining the horizon for humankind's foray into the cosmos.

As we stand on the precipice of a new dawn in space exploration, the role of private enterprises continues to rival, supplement, and even outpace that of public agencies. The journeys of these pioneering companies have not just democratized access to space, but have also equipped humanity with a transformative view of how we regard our own existence in this vast and endless expanse.

It was a journey begun with the loftiest of ideals, dreams shared by luminaries who eyed the cosmos not as an insurmountable wonder, but as a frontier eager for exploration. The dawn of private space exploration is a testament to human ingenuity, resilience, and our perpetual pursuit of what lies beyond the known. It is a tale of turning starlit dreams into concrete realities, one ground-breaking venture at a time. In this grand cosmic play, the lines blurred between the possible and impossible, transforming the very way we interact with space. And as this tale unfolds, the promise of the cosmos grows ever closer for us all – no longer a distant dream, but an attainable frontier, one centered in the palm of our private ambitions.

Despite the challenges that pervade the journey into the cosmos, the continuous surge of innovation and unflagged human spirit point towards a future that is not only within arm's reach but one that's technologically advanced, commercially viable, and above all, inclusive. As the dawn of our cosmic ambitions transforms into its vibrant day, we inch closer towards an exciting reality - a society that's interstellar, and a species that rightfully belongs to the stars.

Chapter 3. Space Tourism: Commercializing Space Travel

The phrase "space tourism" suggests a realm previously exclusive to astronauts and a glittering coterie of the global elite is inching closer to the general public. As a result, a new genre of tourism is steadily gaining traction, promising to imbue everyday lives with a dash of cosmic magic.

The idea of taking a rewarding, albeit costly, vacation beyond the earthly confines has captivated our collective imagination. With the quicksilver developments currently underway, such vacations could quickly metamorphose from dreams to reality.

3.1. Rapid Strides Toward Commercializing Space Travel

This evolution is owed in large part to the initiatives of private companies vying to create a profitable industry from space tourism. Giants like SpaceX, Blue Origin and Virgin Galactic are making it increasingly evident that, if financial and safety considerations permit, space tourism will soon be as commonplace as an international flight.

SpaceX, helmed by technology and business magnate Elon Musk, has been taking giant strides into the cosmos. The company's reusable spacecraft, designed to ferry passengers to orbit and return them safely, is making space travel increasingly feasible. SpaceX's audacious goal: to facilitate life on other planets.

Blue Origin and Virgin Galactic, captained by fellow billionaires Jeff

Bezos and Richard Branson respectively, similarly aspire to commercialize space travel, betting on suborbital Space flights that orbit the earth without circling it. Passengers on such flights experience zero gravity and are treated to a stunning view of our home planet suspended in the inky void of space.

3.2. The Economics of Space Tourism

Beyond the surreal sightseeing experience, space tourism has substantial financial incentives. In a nascent market with massive potential, the economics of space tourism is a fascinating study.

The first component of the economics is the cost—primarily rocket manufacture and launch. These can be prohibitively high, and thus pose the greatest challenge to private companies that cannot expect returns anytime soon. For instance, SpaceX's Falcon 9 launch reportedly costs around $62 million.

However, rapid technological advancements and the use of reusable rockets have led to a drastic reduction in the costs associated. Meanwhile, commercial competition encourages innovation, invariably leading to improvement in cost efficiency.

The second significant aspect to consider when diving into the financial labyrinth of space tourism is demand. Currently, demand for space tourism exists among the uber-rich, a multitude of researchers, adventurous travelers, and even national governments. With lowering costs and increased accessibility, demand could surge exponentially among the wider public keen on an out-of-the-world vacation.

The forecasted revenue of the space tourism market speaks volumes about its potential. According to some estimates, this industry is expected to be worth over $3 billion by 2030.

3.3. Risks and Challenges in Commercializing Space Travel

While the progress and potential of space tourism cannot be disputed, it is important to maintain a balanced perspective by considering the risks and challenges associated with commercializing space travel.

Space journeys are fraught with a plethora of hazards. The harsh environmental conditions of space, the dense technological intricacies involved in a space flight, and the unpredictable nature of space weather together impinge upon safe space travel. Until more research and testing is conducted, space tourism remains a risky proposition.

The other significant challenge comes in the form of governmental regulations. As space tourism essentially entails international and interplanetary travel, regulatory concerns would likely heap up. There are currently no international laws that effectively govern space tourism.

Additionally, there is the continued challenge of cost. Despite the lowering of costs, it's still not a viable vacation option for the majority of people. At its current stages, space tourism is only within the financial reach of the super-wealthy.

3.4. The Future of Space Tourism: So Near, Yet So Far?

Despite the stumbling blocks, the future of space tourism appears promising. As technology continues to evolve and mature, it is realistic to assume that space tourism will become safer, less regulated and more affordable.

It's not inconceivable that within the next couple of decades, holidaymakers will have the choice between sunbathing in Hawaii or experiencing weightlessness in the vast vacuum of space.

Notwithstanding, we must remember to approach our planetary playground with respect and responsibility, so we can preserve the majesty of space for generations to come.

Best described as a thrilling jaunt, space tourism holds extraordinary potential that, if harnessed prudently, could redefine not just tourism, but human life in its entirety. Upon the remarkable canvas of the cosmos, humanity's aspirations and explorations paint a spectacular portrait of courage, curiosity, and consumption. And so, pensively and excitedly, we step into this brave new world—one with an eye to the telescope and a foot on a rocket readying to conquer the untamed universe.

With this exciting venture on the brink, it seems accurate to say that the future is closer than we think. The field of space tourism, once an exotic side thought, may soon become an everyday topic of conversation. However, as we hover on the precipice of this new age, only time will truly tell how we will grasp the potential of space as the newest frontier.

Chapter 4. Technological Advancements Revitalizing the Industry

In the beginning, humanity looked up at the stars with fascination – an elusive fascination that lay just beyond reach. Today, the incorporation of cutting-edge technology is making space travel an accessible dream, invigorating an industry previously monopolized by governmental entities. As we delve into an analysis of the technological developments bolstering this revolution, it is crucial to remember how deeply rooted they are in human innovation and curiosity.

4.1. Trekking the Road to Accessibility

The voyage towards making space tourism viable has been underpinned by some astonishing breakthroughs, the most pivotal of these being reusable rockets. Traditionally, rockets have been one-use constructs. After their mission, they would either burn up in Earth's atmosphere or plummet into the ocean, becoming a sunk cost, both metaphorically and literally. However when SpaceX, led by visionary entrepreneur Elon Musk, introduced the Falcon 1, the trend towards disposable rockets began its shift. The Falcon 1 incorporated a landing mechanism into their design, and in 2008, the first successful orbital launch of a privately-funded, liquid-propellant rocket was achieved, heralding a new era in space exploration.

In successive years, SpaceX has significantly improved their design, developing the Falcon 9 and Falcon Heavy – rockets characterized by their reusability. The first stage of Falcon 9, which bears the main engines and majority of fuel, manages to return to a landing site after delivering its payload to space, marking a monumental achievement for sustainable and cost-effective space travel.

What makes reusable rockets so revolutionary is their potential to drastically cut the costs of access to space. Without the need to construct a new rocket for every launch, costs can be significantly reduced, setting the stage for the prospect of affordable space tourism.

4.2. The Inception and Evolution of Spaceships

While rockets are responsible for transporting payloads to space, it is the space craft or spaceship that holds much of the potential for human-ingenuity. Virgin Galactic's SpaceShipTwo, for instance, captures the essence of this ingenuity in its design. The vehicle, air-launched from beneath a carrier airplane, uses hybrid rocket motors to shoot into space. After reaching its planned altitude, it re-enters the atmosphere in a 'feathering' configuration that involves rotating the wings upwards to increase drag and stabilize descent.

Another remarkable marvel in the development of spaceships is SpaceX's Starship. Designed for fully reusable interplanetary transport, it is likely to replace their Falcon and Dragon crafts in the near future. Once operational, the two-stage spacecraft will have the capacity to transport up to 100 people to Mars, posing myriad possibilities for space tourism, amongst other ventures.

4.3. Advent of Space Stations and Hotels

The conversation about space tourism would be incomplete without discussing the dramatic progress in space habitats. Axiom Space is firm in its ambition to build the first commercial space station, which is expected to serve as a luxury hotel for space tourists – a truly extraordinary feat. Projected to launch in 2024, Axiom Space Station aims to enable tourists to experience microgravity, gyms, research labs, and observation windows with views of Earth that are truly out of this world.

In a similar vein, Orion Span seeks to launch its space hotel, "Aurora Station", into low earth orbit. The hotel promises a full experience of astronaut life, with spacewalks and participation in research experiments amongst the available activities.

4.4. Exciting Innovations on the Horizon

Also of note are the advancements in propulsion systems, materials science, and artificial intelligence, which have significant roles to play in the future of space tourism. For instance, the exploration of nuclear propulsion and ion drives could drastically shorten the travel duration to other planets. Simultaneously, advancements in materials science are leading to lightweight and stronger spacecraft constructions, offering improved safety and capability.

In a similar vein, innovations in AI and robotics are ushering in a new era of space exploration. These technologies have the potential to manage spacecraft autonomously, survey alien terrains, and even perform complex space tasks that are currently executed by highly trained astronauts.

As we stand on the brink of this new frontier, the potential is boundless. New technology is constantly being developed and matured, gradually overcoming the challenges of cost, safety, and sustainability that have constrained progress thus far. The realm of space was once the exclusive domain of superpowers and their astronauts; but thanks to technological progress, this monopoly is shifting, setting humanity on course for the era of space tourism. The dream of becoming an astronaut, once confined to childhood and tempered by age, is becoming an ever increasingly likely possibility.

Chapter 5. The Key Players: SpaceX, Blue Origin, Virgin Galactic, and more

In the ambitious realm of private space exploration and tourism, several companies stand out as trailblazers of the new frontier. Crafting state-of-the-art rockets and spacecraft, SpaceX, Blue Origin, and Virgin Galactic are among the aerospace companies leading humanity into a new era of space access and exploration.

5.1. SpaceX

Space Exploration Technologies Corp., better known as SpaceX, was founded by Tesla CEO Elon Musk in 2002 with a grand vision: to make life multiplanetary. Over the years, SpaceX has revolutionized the aerospace industry, pioneering reusable rockets that can return to Earth and land vertically, a feat that reduces the cost of space launches significantly.

The Falcon and Starship rockets, alongside Dragon crew and cargo spacecraft, are the key technological elements that drive SpaceX's operations. The Falcon 9, Falcon Heavy, and the future-generation Starship are all part of SpaceX's plan for the colonization of Mars.

The Dragon spacecraft, Falcon rockets, and the ambitious Starship serve dual purposes for SpaceX: delivering cargo and humans to space for NASA, the International Space Station (ISS), and other clients, plus facilitating Musk's dream of Mars colonization.

SpaceX has already achieved multiple significant milestones, from launching, orbiting, and recovering the first privately-funded spacecraft (Dragon, 2010) to facilitating NASA astronaut missions from American soil after the Space Shuttle program was retired.

Musk's bold vision of colonizing Mars underpins many of SpaceX's endeavors, pushing the boundaries of what private space companies can accomplish.

5.2. Blue Origin

Founded by Jeff Bezos, the founder of Amazon, Blue Origin operates on the principle of 'step by step, ferociously,' reflected in the company's incremental approach to space travel. Since its founding in 2000, Blue Origin has been developing technologies to enable human access to space and the sustainable colonization of our solar system.

The New Shepard, named after the first American astronaut in space, Alan Shepard, is Blue Origin's suborbital rocket system designed for space tourism. This reusable vehicle is intended to take astronauts and research payloads past the Kármán line – the internationally recognized boundary of space – and return them safely to Earth.

Blue Origin's larger launch vehicle, the New Glenn, named after the first American astronaut to orbit earth, John Glenn, is a reusable heavy-lift launch vehicle designed for payloads and future crewed missions. New Glenn is anticipated to be a significant participant in Earth's growing satellite industry.

Blue Origin also has an ambitious vision for colonization beyond Earth. Project Artemis (not to be confused with NASA's Artemis program) involves developing infrastructure on the Moon to sustain a human colony, transforming humanity into a space-faring civilization.

5.3. Virgin Galactic

Virgin Galactic, founded by Sir Richard Branson, is a company built on the dream of commercial spaceliner service. Its suborbital spaceplane, SpaceShipTwo (VSS Unity), mounted beneath a carrier aircraft known as WhiteKnightTwo, is at the forefront of these efforts.

Unlike SpaceX and Blue Origin, which employ rocket systems for launching payloads into space, Virgin Galactic's approach for space tourism is a little different. The dual fuselage of WhiteKnightTwo carries the SpaceShipTwo vehicle to a high altitude, after which the vehicle detaches and its rocket engine propels it further up to the edge of space.

Virgin Galactic's strategy to space tourism emphasizes not only transportation but also a unique and high-quality astronaut experience. This includes special astronaut suits, luxurious reclining seats, and large windows for a panoramic view of space.

5.4. Beyond the Big Three

Our exploration of space won't be led solely by these three powerhouses. Several other players, including Boeing, Lockheed Martin, Northrop Grumman, and emerging companies like Rocket Lab and Firefly Aerospace, are moving with determination in the field of private space exploration and tourism.

5.5. A Galactic Conclusion

As "The New Space Race" for private space exploration and tourism advances, these enterprises and their unique approaches to conquering space become increasingly important. Rapid technological advancements, coupled with ambitious visions of multi-planetary human life, mean that our acquaintance with the cosmos is set to rise spectacularly in the coming decades. Humanity's giant leap into the era of space tourism and exploration will change not only how we view space but also how we understand our place within it. Essentially, as the key players stride ahead, the final frontier may not be much of a frontier after all.

Chapter 6. Mission Safety and Ethical Considerations

The study of outer space and the effort to inhabit it raises numerous safety and ethical quandaries, which are the subjects we delve into for this discourse.

6.1. Human Safety Considerations

As space tourism seeks to democratize access to space, paying utmost attention to the safety of all future astronauts becomes paramount. The complexities of space travel, from launch and re-entry to survival in outer space, expose humans to extremes beyond their evolutionary design. Each phase of a space mission has unique hazards.

Ensuring human safety involves rigorous testing and redundancy in the spacecraft's engineering design. Adding to it the need for comprehensive pre-flight training for passengers, including emergency response and health check-ups, ensures an additional layer of safety.

Developing safe space vehicles necessitates minimizing the risk associated with human-rating spacecraft, which entails designing to stringent safety standards. These standards encompass an extensive range of objectives, such as preventing astronaut injury, providing a survivable environment in the event of system failures or accidents, and ensuring safe return to Earth.

Establishing safe operating procedures also plays a crucial role. For instance, during a launch, all safety-critical systems must perform a series of automated health-checks. Failure to pass these checks automatically triggers a launch abort to extricate the crew from potential harm.

6.2. Ethics of Space Tourism

Space tourism, while being a thrilling prospect, also brings ethical considerations to the fore. As we depart from a world where space was strictly domain of trained astronauts to one where anyone with the financial means can venture into space, it's vital to examine the issues that arise.

Economic disparity is one of the profound ethical questions in space tourism. Considering the exorbitant costs of space travel today, the opportunity is effectively limited to the very affluent. As it stands, space tourism risks widening the socio-economic divides on Earth even further. Consequently, initiatives to make space travel more accessible are critical to dismantle these barriers.

Another consideration is the environmental impact. Rockets burn considerable amounts of fuel to resist Earth's gravity. This fuel consumption not only depletes fossil fuels but also contributes to atmospheric pollution. As the number of space tourists grows, so will these environmental concerns. Substantial effort is necessary to develop more sustainable launch methods to mitigate these effects.

6.3. Ethical Considerations in Space Exploration

Space exploration, in its pursuit of knowledge and advancement, poses ethical dilemmas of its own. The foremost among these is the potential contamination of celestial bodies with Earth microbes, also termed as 'forward contamination'. This biocontamination could inadvertently wipe out potential extraterrestrial life and hence needs stringent control efforts.

Consequently, spacecraft exploring other planets, moons, or asteroids follow stringent sterilization procedures to limit the risk of forward contamination. These protocols extend to crewed missions as well, with astronauts undergoing quarantine before their missions.

The opposite concern, that of 'backward contamination', proposes the risk of extraterrestrial organisms contaminating Earth. Return missions, particularly from Mars, emphasize spacecraft sterilization and quarantine protocols to safeguard Earth's ecosphere.

In addition, space laws and treaties need to address property rights and territorial claims on celestial bodies. As private enterprises venture into space, they must uphold non-appropriation principles enshrined in the Outer Space Treaty.

Other ethical considerations include preserving the historical value of lunar artifacts or other evidence of human lunar presence. Preserving space heritage promotes scientific study and fuels curiosity about our shared past.

In conclusion, space tourism and private space exploration examine Humanity's capacity to extend its reach while preserving its fundamental values. Balancing safety and ethics in space is a delicate act on this new frontier, but it's a challenge necessary to take up as we step into the future. The cosmos beckons, teeming with opportunities and discoveries, but it's upon humanity to explore it responsibly and ethically. Space is not just a new frontier for exploration, but also a new aquatorium for our ethical and moral development. As we traverse the cosmos, it's vital that we carry our best selves into the stars.

Chapter 7. The Economics of Space Tourism: Costs, Investments, and Returns

Exploring the unbounded cosmos has always been mankind's loftiest aspiration. But until recently, space travel was a privilege reserved for astronauts working for governmental agencies. With eclectic developments in the private space exploration sector, the cosmos's majestic beauty and intriguing mystery are now being unveiled to the public. One key aspect of these organically progressive developments pertains to the economics behind the audacious ventures. It's crucial to understand the various cost verticals, investment landscapes, and expected returns driving this industry.

7.1. Financial Landscape: Cost of Space Tourism

Space tourism ventures necessitate colossal capital expenditures for research and development, technology acquisition, operational services, and safety assurances. The two essential cost components are the non-recurring and recurring costs.

Non-recurring costs cover extensive research, product and infrastructure development, safety system validation, and regulatory compliance. Manufacturers dealing in spacecraft components must invest heavily in research and development, testing, tooling and manufacturing setup, and the production of initial units.

Recurring costs cover the production of additional spacecraft units, operational and maintenance expenditures, crew and passenger training, and mission launch, tracking, and recovery. Additional costs include insurance and marketing. A table providing a typical cost breakdown for a commercial space mission can be illustrative:

```
|Cost Component|Percentage of total cost|
|--------------|------------------------|
|Research and Development|32%|
|Hardware Production|24%|
|Operational Services|15%|
|Safety Validation|13%|
|Regulatory Compliance|8%|
|Insurance and Marketing|8%|
```

The accompanying high ticket prices reflect these costs. A ticket aboard Elon Musk's SpaceX can cost around $55 million, while Virgin Galactic charges approximately $250,000 per flight. On the Russian spacecraft Soyuz, a ticket to the International Space Station (ISS) will set you back a whopping $75-80 million.

7.2. Revenue Streams for Space Tourism Providers

Space tourism providers aren't solely relying on passenger ticket sales for revenue. These private space companies diversify their income streams by providing services such as satellite launches, micro-gravity research platforms, space systems testing, and astronaut training services. Companies like SpaceX use the profits from these activities to subsidize the cost of developing human-rated spacecrafts.

Investing in these alternate revenue streams also allows companies to mitigate risks associated with the nascent and unpredictable market for space tourism. For instance, in 2017, SpaceX had about 60% of the global launch business, but only a small percentage of their revenue was projected to come from space tourism.

7.3. Investment Scenario: Funding Space Ventures

Space tourism, due to its novelty, potential, and inherent risk, attracts a unique blend of investors. The investors can be classified into three categories: billionaire entrepreneurs, venture capitalists, and governments.

Billionaire entrepreneurs, like Elon Musk (SpaceX), Jeff Bezos (Blue Origin), and Richard Branson (Virgin Galactic) invest their own capital. These individuals have a personal interest and ambition in promoting space exploration, thus showing a high tolerance for risk and long-term capital commitment.

Venture capitalists (VCs) have shown increasing interest in the space sector. Investment in space ventures by VCs has been accelerating, with space startups raising $5.8 billion in investment capital globally in 2019, according to Space Capital.

Governments, eager to stimulate the growth of the commercial space sector to augment national space capabilities, provide one of the significant sources of funding as well. For instance, SpaceX has received more than $3.5 billion in contracts from NASA.

Although the potential profits are enticing, space tourism investments are complex, risky, and long-term. Careful due diligence and a balanced approach are critical for investors in this space.

7.4. Returns on Investment: Economic Impact

Direct economic returns for these investments are relatively long term. However, investments in space tourism aren't only about immediate financial profit. Space companies provide indirect returns and influence a wide range of market sectors.

These indirect benefits include driving technological advancements, boosting STEM (Science, Technology, Engineering, and Mathematics) education, fostering innovation, and creating high-paying jobs, thus contributing to economic growth. One study estimates that the space industry could contribute more than $1 trillion to the global economy by 2040.

Moreover, activities like space tourism stimulate the growth of ancillary industries such as spaceports, training facilities, travel agents, hospitality, and merchandising. Each successful launch helps to further reduce costs and makes space tourism more viable for the average person, expanding the market.

As private companies and governments continue to invest in space tourism, the sector promises not only an astonishing new frontier for tourists but also a vast economic landscape to explore. These ventures push human boundaries, create lasting impacts on Earth's economy, and serve as inspiring examples of ambition, progress, and resilience in facing the unknown.

Chapter 8. Regulatory and Policy Challenges

Even as technologies push boundaries and private enterprises boldly take leaps into the outer cosmos, these endeavors are inextricably linked to the regulatory landscape that shapes and navigates this latest frontier. It is a complex weave of international treaties, national policies, laws, and regulatory mechanisms that seek to address the myriad of challenges associated with this new era of exploratory ventures. In this discourse, we will examine the significant regulatory and policy governing space tourism and private space exploration, detailing key issues and proposing possible solutions.

8.1. International Space Law

International space law, primarily the Outer Space Treaty of 1967, underlines the fundamental legal and policy framework governing human activities in space, including private space exploration. However, space law was designed for a time when state-sponsored space agencies were the only entities capable of exploring space. Such law defines outer space as a "province of all mankind" and prohibits national appropriation, thereby raising the question of how the principle applies to non-state actors.

In view of private space exploration, the core challenges lie in the interpretation and application of its somewhat nebulous provisions. As Article VI of the Outer Space Treaty holds, the activities of non-governmental entities require "authorization and continuing supervision" by the state. Opinions differ on what "continuing supervision" entails, especially with evolving technological advancements and the growing commercialization of space activities. The legal lacuna creates an ambiguity that can potentially undermine the credibility, predictability, and sustainability of private space activities.

8.2. National Space Law

On a national level, the United States, through the Commercial Space Launch Amendments Act 2004 and the U.S. Commercial Space Launch Competitiveness Act 2015, has made strides toward creating a regulatory environment conducive for the growth of space tourism and private space exploration. It has sanctioned asteroid mining and provided a legislative framework to facilitate the commercial exploitation of space resources. However, the interplay with international law and the lack of corresponding legal structures in other countries can lead to potential conflicts and regulatory disparities.

The challenge here involves establishing a balanced regulatory environment that promotes innovation while ensuring safety and sustainability. Over-regulation might discourage commercial ventures, while under-regulation risks safety and the potential for reaping long-term benefits from space resources.

8.3. Space Traffic Management

With increasing number of satellites, debris, and possible influx of space tourists, space traffic management (STM) is paramount. The absence of an all-encompassing policy or legal framework to regulate space traffic contributes to the growing problem of congestion and space debris. A collision in space can have a domino effect, known as the Kessler Syndrome, creating significant physical and legal fallout.

The critical concern here is to establish effective STM policies and mechanisms. Ideas range from expanding the 1972 Convention on International Liability for Damage Caused by Space Objects to include guidelines for debris removal, to creating multilateral or even global agreements on STM.

8.4. Environmental Concerns

The environmental impact of increased space activities, both in terms of terrestrial pollution from launches and the potential degradation of celestial bodies, needs strict regulation. The policy challenge is to balance the growth of the industry with sustainable practices.

Ongoing debates center on the lack of a comprehensive legal and policy framework addressing the environmental aspects of space exploration and tourism. This includes laws on planetary protection, preservation of "dark skies", and disposition of nuclear power sources in space.

8.5. Legal Status and Rights of Space Tourists

Finally, as space becomes more accessible, tackling the legal status and rights of space tourists is essential. Existing law treats astronauts as "envoys of mankind", a status unlikely to extend to tourists or private explorers who lack professional training and responsibility. Key questions arise such as the determination of liability in case of accidents, insurance, rescue operations, and even the basic rights and obligations of space tourists.

In summary, space tourism and private space exploration present a regulatory conundrum. The need of the hour, therefore, is a comprehensive and nuanced approach addressing all these aspects. There is an urgency for international discourse and consensus, modification of existing laws, and drafting of new ones that accommodate the evolving realities of outer space exploration. To truly open up the final frontier, we must first navigate the expansive landscape of policy and regulatory challenges here on Earth.

Chapter 9. Impacts on Earth: Environmental and Societal Implications

Space travel has long been viewed as a testament to human innovation and adventurous spirit. Yet, as pioneering space companies aim to bring about a new era of commercialized space travel, we must grapple with the potential consequences. This chapter embarks on an investigative journey into the environmental and societal implications of private space exploration and tourism, assessing the impacts on our beloved home planet, Earth.

9.1. Potential Environmental Impact

Private space travel may usher in a new era of discovery, but it must still confront a daunting challenge: the environmental impact on Earth. Rocket launches consume formidable amounts of energy and yield significant emissions.

The chemical equation for the combustion of rocket fuels, often liquid hydrogen (LH2) and liquid oxygen (LOX) in modern systems, yields water vapor as a major by-product. While this might appear benign, when released into the stratosphere or above, water vapor behaves as a potent greenhouse gas. The crux of the matter lies in the fact that typical aircraft, for example, release their emissions in the troposphere, where water vapor has a short lifespan. Rockets, on the other hand, deposit their emissions directly into the stratosphere, where the water vapor can linger for years and exacerbate global warming.

Aside from direct emissions, rocket launches also generate a particular class of solid particles, known as black carbon particles. These particles can linger in the stratosphere and adversely affect the ozone layer. With continuous launches, the cumulative effect could spell serious trouble for our environment.

Moreover, the construction of infrastructure to support a burgeoning space industry also carries with it increased draw on resources. The activities associated with building and maintaining launch sites, assembly lines, testing facilities, and the like all contribute to environmental degradation.

9.2. Societal Implications

Space tourism promises adventure to individuals beyond previous limits, but it also beckons profound societal implications. Such a shift would exacerbate existing socio-economic divisions, championing a new kind of elite distinction: space tourists.

The ticket to space is currently beyond the economic reach of the average world citizen. Currently, leisure space travel is an exclusive privilege of the extraordinarily affluent. With publicly disclosed prices from companies such as Virgin Galactic starting from $250,000 per seat, it is clear that space tourism will not be egalitarian.

In an era where millions still grapple with fundamental issues such as poverty, education, and health care, the emergence of space tourism draws sharp attention to social inequalities. There is the concern that resources directed toward frivolous space travel could detract from efforts to address terrestrial problems. There remains an ethical question: Is it right to pour billions into space tourism when many deep-rooted socio-economic problems persist on Earth?

Another socio-cultural implication arises from the potential shift in humanity's worldview. As more humans personally witness the fragility and isolation of Earth in the vast cosmos - often referred to as the 'Overview Effect - this might steer collective consciousness toward a more global and ecological perspective.

On the other hand, the commercialization of space raises concerns regarding the commodification of the heavenly bodies. Considering that only a few private entities control space tourism, pressing questions emerge related to monopolization, regulation, and the very ethics of allowing profit-driven corporations to leverage the cosmos for economic gains.

Although space tourism and private space exploration bring promise of advancing human knowledge and catalyzing innovation, it's critically important we assess the potential repercussions on Earth. An eco-centric and socially responsible approach is vital to ensure we balance the scales of progress with the health of our planet and societal well-being. The dawn of the new space age beckons us not only to look up and out but also down and inward, probing our responsibilities as stewards of this pale blue dot we call home.

Chapter 10. The Future of Space Tourism and Private Space Exploration

Over the last few decades, private companies have gradually started to stake their claim amidst the stars, initiating a paradigm shift. For many years, the domain of space was exclusively in the hands of government agencies. Not anymore. The balance of power is shifting rapidly as the era of private space exploration gains momentum.

This privatization of space exploration is fueling a whole new industry: Space Tourism. Now that we are on the precipice of making it a reality, we find ourselves asking the question; what does the future hold for space tourism and private space exploration?

10.1. The Dawn of the Private Era

Let's take a moment to remember when SpaceX, a private company, launched the Crew Dragon in May 2020, transporting NASA astronauts to the International Space Station (ISS). This remarkable feat became a historic moment, earmarking the dawn of an era; a new age of space exploration led by private entities.

As private companies push the bar higher, governments see an opportunity to capitalize on these advances to further their scientific explorations too. An example is NASA's Commercial Crew Program. It is a radical evolution for NASA, which now contracts private companies to journey to the ISS. SpaceX and Boeing are two contracted companies, each developing spacecraft for this purpose.

This transitioning phase marks an inevitable rise of an innovative market economy that will unfold within the cosmos. Major players are already drafting exciting business models, keeping an eye on space's commercial opportunities.

10.2. Bridging the Gap: Space Tourism

From suborbital to orbital tourism, private entities are chipping away at barriers previously deemed impenetrable. Companies like Virgin Galactic, Blue Origin, and SpaceX all give new meaning to the phrase 'the sky's the limit.' Each company, developing cutting-edge spacecraft, intends to make space travel as regular as commercial air travel one day.

Virgin Galactic, for example, offers suborbital tourist flights aboard SpaceShipTwo. The ship is not designed to circle the Earth but to offer a brief taste of space with a few minutes of microgravity and a spectacular view.

Blue Origin's New Shepard spacecraft also offers suborbital space tourism, providing passengers with a few minutes of weightlessness and a look at Earth from space.

Meanwhile, SpaceX plans to go much further—literally—with ambitions to send tourists on multiple-day trips around the moon in their Starship spacecraft.

Yet, space tourism is still at an embryonic stage, and many practicalities and challenges must be addressed for it to mature into a full-fledged industry.

10.3. Economics of Space Tourism

Determining the right price for a ticket to space is no small feat. The current pricing strategies are predictive based largely on launching cost, spacecraft capacity, and exploratory experience.

For now, the pricing is aimed at affluent adventurers. For example, Virgin Galactic's current ticket price for future space flights is at $250,000. The price aimed by Blue Origin is undisclosed but rumored to be in a similar spectrum.

To make space tourism ubiquitous, companies need to significantly reduce the launching costs. One promising way is through rocket reusability, a primary focus of SpaceX.

10.4. Challenges Ahead: Safety and Regulation

As the industry grows, it is crucial to ensure the safety of passengers and to establish a clear regulatory framework.

Space weather, including solar flares and cosmic rays, can pose significant risks. Also, the physical demands of space travel are immense, as highlighted by the rigorous training astronauts undergo.

Covering the legal aspects, the Outer Space Treaty of 1967 provides some regulations, but these may need to be updated or supplemented due to the latest advancements in space tourism.

10.5. The Larger Impact

As space tourism becomes more accessible, it will affect numerous sectors. It's likely to accelerate research in fields like astronomy, biology, and ecology, with potential implications for climate modeling and the assessment of Earth's ecosystems.

Moreover, space tourism could give rise to new educational and career opportunities, inspiring the youth towards science, technology, and adventure.

10.6. Looking Forward

The future of space tourism is promising. The model is inevitably green, focusing on sustainable fuel sources and reusability. This shift towards a sustainable approach could heavily influence the broader aviation industry.

Moreover, the continuation of international collaboration coupled with increased private sector involvement could lead to unprecedented advancements in the industry.

Space tourism is a new industry in a new market with endless possibilities. Only time will reveal how it shapes our planet, society, and civilization in the times to come. As we stand looking into the distant stars, we can almost feel our species setting foot into the cosmos. This indeed is the dawn of the era of private space exploration and tourism, an era filled with unprecedented opportunities and challenges. The future truly is closer than we think.

Chapter 11. Catching Stardust: Potential Opportunities and Unforseen Challenges

The cosmos, with its raw and endless beauty, holds the promise of a future filled with astounding prospects, as well as unforeseen challenges. The private space industry is the new vessel leading us to these potential opportunities, navigating through a vast ocean of stardust, with the objective of advancing knowledge, pushing boundaries, and creating an entirely new market for tourism and exploration.

11.1. Uncovering the Prospects

First and foremost, the potential opportunities that private space exploration offers stand at the forefront of this grand venture. For existing space agencies, private companies' interest and investment bring much-needed financial support. Growth in the private sector space industry may also drive technological advances at a faster rate than state-funded projects, as competition fosters rapid development and innovation. This could range from advanced propulsion methods aimed at reducing travel time, space mining techniques to enhance resource availability, or the creation of sustainable life-support systems for long-term space habitation.

The economic value of these ventures is a potent motivator. The untapped resources of asteroids and planets offer bountiful opportunities. For instance, asteroids rich in precious metals like platinum could be worth trillions. According to industry estimates, this could turn space mining from science fiction to a lucrative commercial enterprise.

Furthermore, private space exploration has the potential to ignite a paradigm shift in the way we perceive our planet. By enabling an increasing number of people to journey to zthe cosmos, we will foster a more universal perspective. This 'overview effect' will transcend boundaries, inspiring a shift towards global unity and sustainable use of Earth's resources.

11.2. Unforeseen Challenges Await

However, this isn't a route laden only with opportunities. Unknown factors and challenges inevitably accompany every pioneering venture pushing towards the unexplored. For example, space travel harbors significant physical risks, from exposure to harmful radiation to the long-term effects of microgravity on the human body. Then there's the daunting task of dealing with isolation – a mental challenge that's often as great as any physical barrier.

The technical obstacles are significant as well. The reliability of components and redundancy systems is a critical concern. Current technology, while impressive, is far from flawless. A small malfunction could result in mission failure, or worse, loss of life.

Moreover, the prospect of high-volume space tourism and exploration opens up ethical and legal challenges that have not yet been adequately addressed. How do we regulate traffic in outer space? What laws apply in the case of accidents? Who has the right to exploit space resources? These are among the many quandaries that need to be audited as we venture beyond our celestial borders.

11.3. Technological Solutions on the Horizon

Despite these challenges, advances in technology and engineering solutions are continuously unfurling to meet them. Radiation shielding materials and more effective propulsion methods are being developed. Extensive research is also underway to understand how space travel affects the human body and devise methods to counteract negative effects.

In terms of hardware reliability, artificial intelligence and machine learning could play a crucial role in making missions safer. Predictive algorithms could help identify potential system failures before they occur, enabling preventative action.

However, it is technology alone that will solve these problems. The human element - adaptability, creativity, and ingenuity - is the linchpin for moving beyond these hurdles.

11.4. The Future: A Mix of Hope and Uncertainty

Thus, as humanity makes strides towards its next frontier, the stars, we find ourselves alongside an exciting blend of promise and unpredictability. What lies beyond our planet holds rich opportunities - scientific, economic, and philosophical. However, the challenges are undeniably great - straining our bodies and minds, testing our technology and our philosophic constructs.

Private space exploration and the flowering of space tourism presents both great potential and unforeseen challenges alike. We are poised at the precipice of a new era, with the universe in front of us. And as we stand at the edge, ready to take this leap, we enter as much into a journey of self-discovery as of cosmic exploration. The stardust that floats through space may likewise trigger a sea of change on a distant blue planet, inciting a revolution of thought, perspective, and existence.

No matter the perils and difficulties that await us in this vast celestial unknown, the allure of the cosmos will continue to seduce us -

strengthening our resolve and igniting our passion to explore. Indeed, this could be humanity's defining moment – a testament to our insatiable curiosity and indomitable spirit. The tale of our venture into space will be a testament to our capacity to dream, dare, and explore... our tale among the stardust.